Marcio Luis Fernandes

FOR A MORE HUMAN GEOGRAPHY

Marcio Luis Fernandes

FOR A MORE HUMAN GEOGRAPHY

on humanism in geography

ScienciaScripts

Imprint

Any brand names and product names mentioned in this book are subject to trademark, brand or patent protection and are trademarks or registered trademarks of their respective holders. The use of brand names, product names, common names, trade names, product descriptions etc. even without a particular marking in this work is in no way to be construed to mean that such names may be regarded as unrestricted in respect of trademark and brand protection legislation and could thus be used by anyone.

Cover image: www.ingimage.com

This book is a translation from the original published under ISBN 978-620-2-80694-7.

Publisher:
Sciencia Scripts
is a trademark of
International Book Market Service Ltd., member of OmniScriptum Publishing Group
17 Meldrum Street, Beau Bassin 71504, Mauritius
Printed at: see last page
ISBN: 978-620-3-12742-3

Zugl. / Approved by: The theoretical profits of this book were defended in 2015 in the doctoral thesis on geography entitled "Guaratiba Island: a place unveiled by its residents flowing into the Olympic River".

FOR A MORE HUMAN GEOGRAPHY

MARCIO LUIS FERNANDES
FOR A MORE HUMAN GEOGRAPHY

MARCIO LUIS FERNANDES

RIO DE JANEIRO

2020

SUMMARY

ABOUT THE BOOK

A more human geography means one that reflects on geographical phenomena with the purpose of achieving a better understanding of man and his condition. In this sense, the humanist approach to geography, with existential phenomenology as the underlying philosophy and hermeneutics as the method of interpretation, seeks to value the experience of the individual or group, in order to understand the mosaic of feelings and the understanding of people in relation to their places. In this perspective, the geographers of the current on canvas argue that their approach deserves the label of humanist, since they study aspects of man such as meanings, values, goals and purposes, as well as allegories, dreams, reveries and reminiscences. This book discusses the assumptions of the humanist horizon, establishing this bias as a proposal for the humanization of geography.

KEY WORDS: Humanist geography. Hermeneutics. Phenomenology. Space. Place.

FIRST WORDS

In Lowenthal's conception (1982, p. 137), a memorable geography would not be tied to mere compendium texts - much less to generalizing perspectives that ignore particularities, individualities and singularities - "[...] but to interpretative studies incorporating a strong personal point of view". Following this premise, we understand that a geography committed to universalizing aspects that embrace society as a whole, excluding the particular universes of individuals and social groups, cannot account for a genuinely human and personal geography forged by each informal geographer in his or her lived world (LOWENTHAL, 1982; COSGROVE, 2004).

By despising the rich material represented by the experiences of the lived world, as well as the conceptions that derive from these experiences, "geography has gone a long way to introduce man as a thinking being in his research" (MELLO, 1991, p. 1). In the early 1970s, however, some geographers frustrated with a geography where the human being represented only one more element of the studied landscape (MORAES, 2007), "began to search in the philosophies of meaning answers to their anguishes" and paths to break with the positivist and neo-positivist assumptions that prevailed in geographic science (MELLO, 1990, p. 22). Surgia a geografia humanista, uma perspectiva que focalizes o homem no centro de todas as coisas, vez que todo ser humano pensa e filosofa, sendo portanto capable de refletir sobre as fenômenos do mundo(s) vivido(s) (RELPH, 1976; BUTTIMER, 1982; TUAN, 1982 MELLO, 2000).

This book will seek to present a proposal for the humanization of geography through the assumptions of humanism in geography. Thus, its theoretical bases have been sought in humanistic authors based on subjectivities, symbolisms, and identities, therefore, in followers of the phenomenological method and hermeneutics such as Relph (1976), Buttimer (1982), Tuan (1982; 1983; 1998), Holzer (2001; 2008), and Mello (1990; 1991; 1993; 1999; 2000).

From the process of renewal of geography and the emergence of critical geography - understood as the currents that broke with the positivist and neo-positivist ideals - geographers of various methodological orientations (existentialist, Marxist, eclectic, etc.) began to assume a more human perspective, seeking through the transformation of the social order, a more generous geography and a more just space or place, which is organized according to men (MORAES, 2007).

In view of the above, let us consider the pertinent questions regarding the assumptions of the humanistic horizon, as well as the principles defended by hermeneutics and phenomenology, these being the main philosophical foundations of the current that guides this book.

1   IN SEARCH OF A MORE HUMAN GEOGRAPHY

In Lowenthal's conception (1982, p. 137), a memorable geography would not be tied to mere compendium texts - much less to generalizing perspectives that ignore particularities, individualities and singularities - "but to interpretative studies that incorporate a strong personal point of view. Following this premise, we understand that a geography committed to universalizing aspects that embrace society as a whole, exempting the particular universes of individuals and social groups, cannot account for a genuinely human and personal geography forged by each informal geographer in his or her lived world (LOWENTHAL, 1982; COSGROVE, 2004).

Introdução

A Geografia Humanística reflete sobre os fenômenos geográficos com o propósito de alcançar melhor entendimento do homem e de sua condição.

A Geografia Humanística procura um entendimento do mundo humano através do estudo;das relações das pessoas com a natureza,do seu comportamento geográfico bem como dos seus sentimentos e idéias a respeito do espaço e do lugar.

Temas da geografia humanística

Território e lugares

- As atividades humanas quanto ao território e lugares têm uma clara semelhança, com os dos outros animais;
- Um caso que esclarece a peculiaridade humana é a importância que as pessoas dão aos eventos biológicos do nascimento e da morte;
- Como um mero espaço se torna um lugar intensamente humano? é uma tarefa para o geógrafo humanista.

By despising the rich material represented by the experiences of the lived world, as well as the conceptions that derive from these experiences, "geography has gone a long way to introduce man as a thinking being in his research" (MELLO, 1991, p. 1). In the early 1970s, however, some geographers frustrated with a geography where the human being represented only one more element of the studied landscape (MORAES, 2007), "began to search in the philosophies of meaning answers to their anguishes" and paths to break with the positivist and neo-positivist assumptions that prevailed in geographic science (MELLO, 1990, p. 22). Humanistic geography was born, a perspective that focuses man on the center of all things, since every human being thinks and philosophizes, and is therefore able to reflect on the phenomena of the worlds lived (BUTTIMER, 1982; MELLO, 2000; RELPH, 1976; TUAN, 1982).

A more human geography means one that reflects on geographical phenomena with the purpose of achieving a better understanding of man and his condition (TUAN, 1982). In this sense, the humanistic approach in geography, with existential phenomenology as the underlying philosophy and hermeneutics as the method of interpretation (BUTTIMER, 1982; GOMES, 2007), seeks to value the experience of the individual or social group, in order to understand the mosaic of feelings and the understanding of people in relation to their places. In this perspective, the geographers of the current on canvas argue that their approach deserves the label of humanistic, since they study the aspects of man such as meanings, values, goals and purposes (CHRISTOFOLETTI, 1982), as well as allegories, dreams, reveries and reminiscences (MELLO, 1991; FERNANDES, 2014).

1.1 Another horizon in search of the humanization of geography

The humanistic movement highlights man and treats him with his meanings, values, objectives, dilemmas and actions in opposition to the abstract, mechanistic and deterministic approach of previous paradigms. Criticism of the reductionist vision of man, especially after 1970, favored humanistic geographers the interpretation of feeling and understanding of the relationships between men and their lived world. This perspective, by advocating the incorporation of the subjective dimension and the experience lived by individuals and social groups in their approaches, proposes an understanding of the human world through the study of people's relationships with nature, their geographical behavior, as well as their feelings regarding space and place (TUAN, 1982; 2013).

The above mentioned ideas appear as relevant to this geographical trend. In place, the individual finds himself fit and even integrated. Such conceptual expression composes this world full of feelings and affections, a center of significance or a focus of man's emotional action. The place is not any locality, but one that expresses affectivity and values for the individual or his collectivity (CHRISTOFOLETTI, 1982). In counterpoint, space is represented by any portion of the Earth's surface, being wide, unknown, undifferentiated, rejected or even hated (MELLO, 1990; 1991; 2001; TUAN, 1982; 1998; 2013), possibly to be captured or conquered.

LUGAR

➡ A discussão teórico- metodológica sobre lugar na ciência geográfica tem sido feita em três perspectivas, tendo como objetivo de ultrapassar a idéia desse conceito como simples localização espacial absoluta:
- na geografia humanística lugar é onde se torna familiar ao indivíduo, é o espaço do vivido, do experenciado.
- na concepção histórico- dialética lugar é como o meio de manifestação da globalização.
- Pela ótica do pensamento pós-moderno, coloca em questão de totalidade para a explicação de lugar.

Despite the distinction between these key concepts of geography in general (CORRÊA, 2002) and the humanistic ramification in particular (HOLZER, 2001; 2008), in experience, the meaning of space can merge and/or be confused with that of place. What begins as an undifferentiated space becomes a place as we know it better and endow it with value (TUAN, 2013). According to Mello (1990, p. 105):

> Certain spaces only become places after a long experience. What is initially ugly or even hated, with time it gains places. Spaces become places due to contact with other people and in affective and economic exchanges, etc.

Humanistic geography does not neglect the affective dimension in the midst of the lived/existential. Therefore, it is fundamental for humanistic-inspired geographers not only to approach the "spatial distribution of social facts, but also the way people live in the places they live or visit, drawing an experience from them" (CLAVAL, 2001, p. 46). In this trail, "being together, being close, does not mean physical closeness, but the affective relationship with another person or another place" (CHRISTOFOLETTI, 1982, p. 23). Places and people physically distant can be affectively very close. Therefore, the study of spaces and places goes back to the analysis of the spatial feelings and ideas of people and groups of people (TUAN, 1982). "From a positivist perspective, geography concerns only the analysis of spatial organization. Under the humanistic horizon, space and place assume very different characteristics" (CHRISTOFOLETTI, 1982, p. 23), and it is up to the geographer of the humanistic wing to translate what they represent through a coherent structure (CHRISTOFOLETTI, 1982; TUAN, 1982).

From the valuation of attitudes comes the concern with tastes, preferences, characteristics and particularities of places. We also value "the environmental context and the aspects that result in the enchantment and magic of the universes lived, in their personality and distinction". There is then the interweaving between the person or social group and the place (CHRISTOFOLETTI, 1982, p. 23), since the individual is not distinct from his or her lived world (RELPH, 1976).

Lugar

- Esse conceito era utilizado no senso comum como sinônimo de localização.
- "A geografia é a ciência dos lugares, não dos homens" – Paul Vidal de La Blache
- Geografia quantitativa – estuda a organização espacial e a palavra lugar tinha o sentido de localização
- Geografia humanista - espaços vivenciados pelas pessoas em suas atividades cotidianas de trabalho, lazer, estudo, convivência familiar, etc.
- Milton Santos - espaço produzido por duas lógicas, a saber, a das vivências cotidianas das pessoas e a dos processos econômicos, políticos e sociais que constituem a globalização.

Prof. Claudinei Perencin

Tuan (1982, p. 159), in turn, reports that "the contribution of humanistic geography to science lies in the revelation of materials of which the scientist, confined to his own conceptual bubble, may not be conscious. For the researcher:

> This material includes the nature and range of human experiences and thoughts, the quality and intensity of an emotion, the ambivalence and ambiguity of values and attitudes, the nature and power of the symbol and the characteristics of human events, intentions and aspirations.

To elucidate this mosaic of material and immaterial objects in the midst of subject/object indivisibility, the humanistic geographer:

> It must have a penetrating interest in philosophy, since it raises fundamental questions of epistemology for which we can seek explanations in the real world. Philosophy also provides a unified point of view from which a whole series of human phenomena can be systematically evaluated (TUAN, 1982, p. 161).

In these terms, we are talking about hermeneutics, which according to Mircea Eliade (1971), is the only effective method of interpretation that humanism cannot deprive itself of, and the phenomenology in which humanistic geography seeks elements to guide its research (GOMES, 2007).

1.2 Philosophies of meaning: methodological foundations of humanistic geography

Based on the principles of phenomenology and hermeneutics, humanistic geography is interested in understanding the soul of places from the experiences lived by individuals and social groups. This perspective understands that place is an integral part of being, being each individual an informal geographer able to talk about the soul of places, because it is man who produces, learns, lives and transmits geography (BUTTIMER, 1982; COSGROVE, 2004; LOWENTHAL, 1982; MELLO, 2000; SCHUTZ, 1979).

"Phenomenology is the philosophy present in a larger number of humanistic studies in geography" (MELLO, 1991, p. 36), being also considered a research method (ARANHA, 1996; GOMES, 2007). "Its creator, the German philosopher Edmund Husserl (1859 - 1938) criticizes scientific theories, particularly those of positivist inspiration, excessively attached to objectivity and the belief that reality is reduced to what is perceived by the senses" (MELLO, 1991, p. 36).

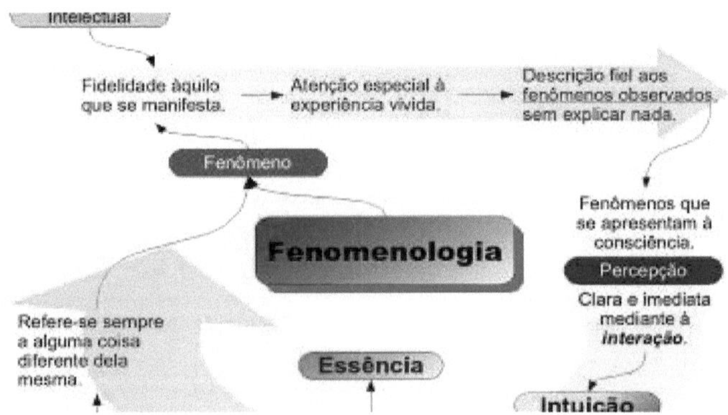

Etymologically, phenomenology is the study of the phenomenon, its role being to analyze the dynamics that provide meaning and meaning to objects, treating the world and human beings in an inseparable way (SCHUTZ, 1979). Examining the concept of phenomenon, which in Greek means what appears (ARANHA, 1996), we understand better that phenomenology deals with knowledge as it appears, that is, as it presents itself to the consciousness. As a source of meaning for the world, the consciousness is not restricted to mere intellectual knowledge, but is the generator of intentionalities that are not only cognitive but also affective and practical. The look on the world is the act by which man experiences it, imagining, judging, loving, fearing (ARANHA, 1996; SCHUTZ, 1979).

FENOMENOLOGIA

Ciência que tem como objetivo descrever tudo aquilo que nos aparece (os fenômenos)

- Não há separação entre a consciência e as coisas.
- Toda consciência é consciência de alguma coisa. (*consciência de...*)
- O mundo e todos os seus entes somente existem porque temos consciência deles.
- Aquilo que não temos consciência não existe.
- Eu sou consciente dela enquanto ela continua existindo fora de mim.
- A consciência não cria os objetos do mundo e também não os percebe como exatamente são.
- A consciência é intencional, ela permite que percebamos o significado das coisas.
- A consciência opera a partir dos fenômenos que ela observa. Por isso, é importante "educar" a consciência a partir do método.

The phenomenology disapproves the naturalistic tendency that guides the human sciences method. For this philosophy, there are no facts with the objectivity intended by positivism, since we do not understand the world as a raw data, devoid of meaning. The world I decode is a universe for me, hence the importance of feeling, understanding and the network of meanings that involve our lived universes (ARANHA, 1996). For Schutz (1979), the irreducible starting point for the phenomenological bases concerns the experiences of the conscious human being, who lives and acts in a world that he captures, interacts and interprets - assuming multiple meanings.

5 teses para entender a Fenomenologia

- Tudo o que é dado à consciência é fenômeno.
- A consciência não é um vazio que será preenchido.
- A consciência não se separa do Ser.
- A consciência não tem um dentro: ela é o "fora" de si mesma.
- A consciência é, sempre, a intenção de alguma coisa.
- Não há divisão entre eu e mundo. O Ser é um ichloss (sem-eu).

Everything that is experienced by human beings corresponds to experiences in/from, their lived world. They constitute it, are directed to it, are tested and experienced in it. The lived universe is simply the entire sphere of everyday experiences, directions and actions through which individuals deal with their interests. According to phenomenological philosophy, each individual builds his own world. Thus, subjectively, two people could never experience the same situation in the same way (SCHUTZ, 1979). In this regard, Buttimer (1982, p. 167) points out that:

> Phenomenologists have been the most systematic spokesmen in a combined effort to reconcile heart and mind, knowledge and action in our daily worlds. Challenging many of the premises and procedures of positive science, they have exposed a radical critique of reductionism, rationality, and the separation of subjects and objects in empirical research. With the existentialists, they preach the argument of liberation from lived experience, appealing for more concrete descriptions of space and time, and their meanings in daily human life.

By contemplating as a common trait the subject-object inseparability, phenomenology examines in a radical way the phenomena of consciousness or lived experience, searches for the facts as they are produced, interpreting the apprehension of the essence and in this way investigates the acts and the understanding about the lived world. In this context:

> The place emerges as a key concept in humanistic geography arising from the phenomenological notion of a world lived emotionally, modeled, introjected, and covered with events, people, itineraries, fights, ambiguities, entanglements, dreams, outbursts, "songs my mother taught me", territorial basis, and all sorts of elements that allow one to feel at home or, on the other hand, distanced in the midst of a topophobic strangeness (MELLO, 2005, p. 34).

> For a long time, geographers have excluded from their approaches neighborhood ties, the stock of knowledge, pleasantness, topophobia, attachment to spaces and places, daily experiences and the links that unite people with the environment. The phenomenology, considering these attributes, serves as a bridge to these specialists, with a view to understanding the world lived, because - unlike the science that omits the questions of life - it does not treat the world independently of human beings (...). With support in the lived world, the geographer can understand how the magic of places is born, the intrinsic particularities of each territorial portion, the distinction of different points of the city, the enchantment, the contempt, the attraction and what is typical of places (...). The world lived is the consciousness and the intimate environment of each one, emotionally modeled and covered with events, relationships, ambiguities, involvements, values and meanings, which includes human beings with all human action and interests, works and sufferings (MELLO, 1991, pp. 37-38).

> The world lived of each one already existed before the birth of the person, who lives and interprets his world from values and stocks of personal experiences, and also through other individuals who transmit to him past and present knowledge (MELLO, 2000, p. 57).

Inter-subjectivity refers to the world common to different people, scenario and object of human beings' actions and interactions. The lived world, continuously experienced, is modified by human actions, which also modify their actions (MELLO, 1991). "The stock of experiences is a daily, practical and theoretical enrichment that provides man with elements to act and think" (MELLO, 1991, p. 38). "However, this knowledge is not homogeneous, but incoherent, partial, contradictory and ambiguous"

(MELLO, 2000, p. 58). "Knowledge of the world, received by formal and informal culture and completed by personal experience, generates intimacy and affection for the place lived" (MELLO, 2000, p. 58). This world which, for phenomenology is the context within which consciousness is revealed, occurs to our experience and interpretation (BUTTIMER, 1982; ELIADE, 1971; GOMES, 2007; MELLO, 2005).

"Hermeneutics, another philosophy of meaning, used by humanistic geographers, has as its precursor the German Wilhem Dilthey (1833 - 1911) who added to this philosophical movement - close to Schutz's phenomenology - elements of interpretative importance" (MELLO, 1991, p. 41).

hermenêutica

Arte de interpretar os livros sagrados e os textos antigos.
Doutrina ou ciência cujo objetivo se caracteriza na interpretação ou compreensão dos textos de teor religioso ou filosófico: hermenêutica sagrada.

[] Dicio.com.br

The origin of the term lies in antiquity, inspired by the Greek mythology of Hermes, god of communication, in charge of bringing the messages of Olympus (GOMES, 2007). Originally used by ancient theologians as a methodology proper to the interpretation of the Bible, the term later came to designate every effort of scientific interpretation of a difficult text that requires an explanation. Contemporarily, hermeneutics constitutes an interpretative or comprehensive philosophical reflection on symbols, being fundamental in all humanities and in all disciplines that deal with the interpretation of the works of men (ABBAGNANO, 2007; JAPIASSÚ E MARCONDES, 2006; PALMER, 1970).

HERMENÊUTICA

☐ **Teoria ou filosofia da interpretação do significado, do sentido; atualmente tópico essencial na filosofia das ciências sociais, da arte e da linguagem, assim como na crítica literária.**

"In hermeneutics," as in phenomenology, "there is no separation between subject and object. Thus, this "interpretative philosophy tries to explain the contents of the mind and other aspects of lived experience" in the midst of introjections, where individuals and social groups are not dissociated from their experienced territorial base. It is up to the geographer to clarify the meaning of concepts, symbols and aspirations, as all this is tied to space and place. In other words, "the humanistic geographer has the task of interpreting the ambivalence, ambiguity and complexity of the consciousness of individuals and/or social groups regarding the environment" (MELLO, 1991, p. 42).

It   is worth noting that "the boundaries between phenomenology and hermeneutics are not very rigid. Thus, "several humanistic geographers -

among them Tuan, Buttimer, Lowenthal and Relph - although they classify themselves as phenomenologists, exhibit the hermeneutical movement in an unmistakable way" (MELLO, 1991, p. 42).

On the humanist side, it is worth repeating, the spatial distribution of social events and facts does not represent the determining element of the research. What is fundamental for geographers on this horizon is the existential way of life of people in the places where they live or those they visit, drawing experiences from them (CLAVAL, 2001).

1.3   **Decoding concepts and humanistic themes**

Commonly, the concepts of space and place express - metaphorically and respectively - the notions of darkness and luminosity (MELLO, 2000). However, far from the positivist and neo-positivist dictates, there is no pre-established rule for opaque spaces, immersed by penumbra, to reach, through their illumination or clarity, the level of place. Paraphrasing Mello (2003), we can point out that the place can lose or receive such a condition depending on the darkness or the brightness with which it is blunted or illuminated in the course of time. In Tuan (1983, p.179) "place is any stable object that captures our attention. Thus, "many places, highly significant for certain individuals and groups, have little visual notoriety" (TUAN, 1983, p. 180). No matter how opaque or visible they may be, certain objects or places that are admired by one person may not be noticed by another (TUAN, 1983).

For Tuan (1983, p. 184), the place is also "a reservoir of memories and dreams. In the meantime, visual notoriety alone is not a guarantee that certain spaces will become places. The deep sense of place results from a combination of historical, cultural, economic, locational, existential, subjective, intersubjective, invisible and visual factors and values (YÁZIGI, 2003). As for the line of thought undertaken here, visibility would not be tied only to human constructions and to the elements of nature that confer visual values to certain fixed ones. However, there is no denying that these cultural and natural attributes, in many cases, represent factors of transformation of certain localities that, with this, may suffer qualitative or non-qualitative changes, depending on the different perspectives of their experiencers. In relation to the differentiated viewpoint of individuals in relation to visibility, Tuan (1983, p. 184) stresses that "[...] most places are not deliberate creations, since they are built to satisfy practical needs". In this context, they acquire visibility and meaning for both local and foreign inhabitants (TUAN, 1983).

In Gomes' conception (2007), the first fundamental characteristic of humanism taken up by geography concerns the unavoidable anthropocentric vision, according to which man is the measure of all things. Moreover, human beings, individually or in groups, tend to place their lived place as the center of the world. In this context, egocentrism and ethnocentrism become universal human traits (TUAN, 1980). Based on Tuan (1980), Mello (1991, p.202) describes ethnocentrism as:

> A universal phenomenon of overvaluation of the "center", "navel", "healthier" or "better place in the world" and can also be understood as collective egocentrism. The people of the "center" establish discrimination between "us" ("superiors") and "them" ("of lesser value", "of inferior culture") by looking at them in a "blasé" way and, at times, with apathy, sarcasm or aggressiveness.

The notion of center is one of the most relevant cultural manifestations, since people commonly tend to place the place where they live as the most important and favorable, and as the center of their world. "All ancient peoples place themselves as the center of relationships and organize what they understand by world in this reference. In this sense, the top is a correlated notion. "A whole religious and cartographic symbolism derives from this centrality, organizing the conception and geographical relationship of these peoples" (MOREIRA, 2009, p. 67). According to Indian beliefs, Mount Meru would be erected in the center of the world. An Iranian belief states that the sacred mountain of Elburs would be located in the center of the Earth. The name of Mount Tabor, in Palestine, could mean "navel". Mount Garizim, in central Palestine, undoubtedly enjoyed the prestige of a central place, as it was called the "navel of the Earth" (ELIADE, 2007). Palestine, according to ancient tradition preserved until today in the region, in its condition as the highest country for being near the summit of the cosmic mountain, would not have been covered up by the flood. A rabbinical text states: "The land of Israel was not submerged by the flood. For the Christians,

> Golgotha was situated at the central point of the world, since it was the summit of the cosmic mountain and, at the same time, the place where Adam had been created and buried. Thus the blood of the Savior is poured on Adam's skull, buried precisely at the foot of the Cross, serving for his redemption. The belief that Golgotha would be situated at the center of the world is still preserved in the folklore of Eastern Christians (ELIADE, 2007, p.24).

Through the beliefs we make mention of, we can deduce that each oriental city was located in the center of the world. For some of these people, the highest point of the cosmic mountain would not only be the highest point on earth, but also the navel of the world, the point where creation began. Some traditions explain the symbolism of the center in terms taken from embryology, according to which "the Divine Being created the world as an embryo. Just as the embryo began to pass from the navel, God began to create the world from the navel onwards, and from there it spread in different directions" (ELIADE, 2007, p. 25). In the meantime, the world would have been created from Zion and the universe conceived from a central point. The creation of man would also have happened in a central point, in the center of the world (ELIADE, 2007).

According to Mesopotamian tradition, man would have been formed in the "navel of the earth". Thus Paradise, where Adam was created from clay, is located in the center of the cosmos. Paradise was the navel of the Earth, and, according to a Syrian tradition, it would have been established on a mountain higher than all the others. Adam would have been created in the center of the Earth (ELIADE, 2007). Based on the premise that creation derived from a center, we can also assume that any place founded has its building in the center of the world of the individuals who established it as home, shelter, refuge and abode.

Unlike traditional critical social theory - in which the concept of place is linked to the local sphere (CARLOS, 1996; SANTOS, 2002) - from a humanist perspective, such an aphorism, for not having a defined scale, becomes too diffuse since it can either designate a seat or encompass the whole world (TUAN, 1983). However, the same Tuan that takes up the maxim that defines Geography as "the study of the Earth as the home of people" (TUAN, 1991, p. 89), also points out that "[...] topophilia sounds false when it is manifested by an extensive territory" (TUAN, 1980, p. 116). However, patriotic love, significant in its dimension, can contradict its elucidations. For the above-mentioned geographer, topophilic feelings need "[...] a compact size, reduced to the biological needs of man and the limited capacities of the senses" (TUAN, 1980, p. 116). Moreover, a person can more easily identify with an area if it indicates a natural unit, small enough to be known personally (TUAN, 1980). In this field, the place is confused with the local sphere, being also the locus of daily life "responsible for human passions through communicative action and various manifestations of spontaneity and creativity" (SANTOS, 2002, p. 322).

Every place has multiple dimensions. In order to discover the dynamism in which it is surrounded, its multiplicity must be considered. But what dimensions are we referring to? In search of the answers, let us consider the following elucidations.

2 THE MULTIPLE DIMENSIONS OF LIFE

As we observed in the previous section, the humanistic orientation emerges as a reaction to positivist and neo-positivist assumptions where the separation between subject and object is clear. Moreover, in positive science, quantitative methods are prioritized, and the adepts of these approaches are interested only in materiality, remaining absent from their studies, subjectivity, symbolism, and existential questions linked to the experiences of individuals and social groups (MELLO, 1990; FERNANDES, 2014).

A) Ao revisitar os lugares, no mundo atual, encontraremos novos significados, novas identidades, pois esses são construídos no cotidiano/mundo vivido e considera as variáveis: objetos, ações, técnica, informação e tempo.

B) A realidade que se constrói no lugar é tensa, pois se refere a um dinamismo que se recria a cada instante/momento entre homens, empresas, instituições e meio ambiente construído.

C) No lugar as relações são, permanentemente, instáveis, em que globalização e localização, globalização e fragmentação são termos de uma dialética que se refaz com freqüência.

From the advent of interpretative perspectives, the different meanings inherent to the lived universe of social groups become the focus of geographic research (CORRÊA E ROSENDAHL, 2012). In this context, geographers begin to focus on the bonds that individuals weave between themselves and their lived universe, as well as why men and/or women attribute meaning and subjective values to their places (CLAVAL, 2007).

In Gomes' (2007) words, one of the main characteristics of humanism in geography is its holistic character, glimpsing the totality in order not to lose the richness of the whole. According to Holzer (2008), the idea of a discipline centered on the study of human action and imagination and the objective and subjective analysis of its products, which intended to be a science of synthesis that was beyond the Cartesian and positivist parameters, takes us back to the 1920s. Following this strand, in 1947, Wrigth - through the concept of geosophy - establishes the basis of a science project that encompasses the different approaches as well as the objective and the subjective indistinctly (WRIGTH, 1947; LOWENTHAL, 1982; HOLZER, 1992; CASSIRER, 2011).

Paraphrasing Almeida (2010), in each place there is an objective and another subjective dimension. Both, together, constitute what is lived. The immaterial dimension, however, is that which gives the connotative component that ends up being, equally, an inseparable part of the concrete dimension (ALMEIDA, 2010).

In the course of the process of renewal of geography, both its character of rupture with previous patterns (GOMES, 2007) and attempts to reconcile polarized positions (ALMEIDA, 2010) are common. As Claval (2007) points out, materialistic geographers are interested in spatial structures and humanistic geographers in initiatives and individual and collective actions. For Claval, the increasing integration between the subjective and objective dimensions of the place is an example of analysis in the intermediate route between two approaches: the spatial and the existential; urbanism and humanism; the material and the immaterial, etc.

> *No dicionário: 1. Espaço ocupado ou que pode ser ocupado por um corpo; 6. Ponto de observação.
>
> *Na Geografia: São os pontos que nos são familiares e que fazem parte da nossa vida, é a porção do espaço que é vivido por habitantes, muitas vezes, este é capaz de criar identidade para indivíduos.
>
> *Na Geografia Humanista: É o ponto em que o individuo se encontra ambientado, que esta integrado e que tem significância para o ele.
>
> ***O que é Lugar?**

In turn, Antonello (2010), seeking a synthesis between the objective and subjective dimensions, points to being in the place where the lived experiences are anchored. The author also affirms that geography must seek to overcome the limits imposed by the division of scientific work, seeking to tread a path that leads to the plurality of foci of analysis in order to understand the complex reality that surrounds us (ANTONELLO, 2010).

In fact, Haesbaert (2005), another geographer who advocates the indissociability between objective and subjective aspects, proclaims in his research that territoriality emerges as the fruit of the interweaving between the different dimensions of reality. The author warns that it is precisely by making a rigid separation between territory as domination (material) and territory as appropriation (symbolic) that many ignore the complexity and richness of the multiterritoriality in which we are involved (HAESBAERT, 2005).

In Bastos (1998), the real is not only constituted of the material universe, but also of the meanings coming from the relationship of individuals with their lived world. For the researcher, in the apprehension of geographic space there is a concrete dimension (the production of material space) and a symbolic dimension (the representations). These two dimensions of reality interact with each other in a symbiotic (subject-object) and inseparable relationship. In this sense, spaces and places can be represented according to an imaginary in which their materiality should not be denied. In this trail, it is not justified to separate the subjectivity of the researcher when interpreting the studied place, since it is presented in such a way that the concrete and the abstract are inseparable parts of the same reality. In the dynamic universe where man is the geographical actor and the place his niche of collective belongings and memories, there is an imbrication between the material and the symbolic, between the spatial and the existential, between the objective and the subjective (BASTOS, 1998).

Like the above-mentioned researchers, Cosgrove (2003) proclaims a geography in which the objective and subjective dimensions of reality are considered. For the above-mentioned geographer, human beings promote spatial transformations with their sensory and material reality. In this sense, all human activity is material and symbolic at the same time. Despite being symbolically constituted, the lived world is material and should not deny its objectivity (COSGROVE, 2003). The lived experiences and other subjective experiences of individuals and social groups occur in a context and should not be seen as independent of materiality (CORRÊA, 2003).

According to Geertz (2013), in anthropology - equally - there is an endless debate about whether culture is subjective or objective. For the author, culture must be understood as an intertwined system of interpretable signs (symbols), as a context in which representations and substantive content are interconnected. In relation to this unnecessary complication, related to the dualism between the concrete and the symbolic, Geertz points out that he has tried to resist the unmeasured subjectivism and has tried to maintain the analysis of symbolic forms considering both dimensions of reality (GEERTZ, 2013).

Tuned in with the researchers cited in this section (CORRÊA, 2003; COSGROVE, 2003; BASTOS, 1998; HAESBAERT, 2005; ANTONELLO, 2010; ALMEIDA, 2010; CLAVAL, 2007; WRIGTH, 1947; LOWENTHAL, 1982; GEERTZ, 2013) I believe that a multidimensional approach is possible in geographic studies, where - in the same way - the different facets of the lived universe are considered. Was it not in the gap left by the exacerbated objectivism and the consequent neglect of the subjective values of individuals and social groups, linked to positivist and neo-positivist paradigms, that humanism in geography emerged?

In my conception, the result of the divorce between the objective and subjective dimensions would be the construction of a geography that does not commune with the desires of a science that was born with the ambition of discovering the world in its totality and diversity (SILVA, 1988; CLAVAL, 2007; MARANDOLA JR, 2012).

In the following topics, we will focus on some concepts and themes that can contribute to approaches that privilege the place in its multiple dimensions, starting with the memories attached to the different places.

2.1 About **the memories of the places**

As we have argued in this research, places are assigned many dimensions of meaning (BUTTIMER, 2015). One of the paths pointed out by different authors in order to explore the dynamic and multifaceted universe lived would be through a glimpse of the intimate and particular collections of their experiencers. However, since every lived experience goes back inexorably to the past, if we want to follow this path, we will need to resort to the subsidy of the individual and collective memories of individuals and social groups (LOWENTHAL, 1982; 1985; 1998; ABREU, 1998; MELLO, 2002; HALBWACHS, 2013; BUTTIMER, 2015).

No matter how much a locality transforms as a result of an urbanization march, for example, it will always retain a residual content of previous temporalities. There is, however, a range of events that leave no physical trace. These are often recorded in the memory of the individuals who accompany such changes. What we think and feel in relation to our geographies is obviously based on experiences lived somewhere in the past. From this link, everything that occurs in our experienced ground becomes part of our own existence (LOWENTHAL, 1998). Daily life has the place as its base, being the result of all our moments, the sum of our memories and the product and junction of all our lived experiences (MENDILOW, 1960).

Possessed by the past, Lowenthal admits to living in a tangle of times. For that thinker, the past resides in us through our memories and memories of different moments (LOWENTHAL, 1985; 1998). By populating the thoughts of human beings, the past is alive in our memory (HIGHET, 1949). In fact, the scenarios and experiences glimpsed and experienced become part of our memory (BUTTERFIELD, 1965).

"Every consciousness of the past is founded on memory. Through memories we recover consciousness of previous events, we distinguish yesterday and today, and confirm that we have already lived a past" (LOWENTHAL, 1998, p.75). Memory has a personal and collective character at the same time. The memories of individuals and social groups sustain their sense of identity in relation to their experienced ground. Our life is impregnated by our memories. At every moment, we bring back some event from the past (LOWENTHAL, 1998).

Memories tend to accumulate with age. Our stock of memories increases as life goes on and experiences multiply. Although there are individual and collective memories, memory is always personal. Memories cease to be personal only when we decide to share them (LOWENTHAL, 1998). The memory of the place is inscribed, not only in some artifacts of the past that persist in its landscape as geographical testimonies, but - mainly - in the memories of the experiences lived by men and women in their lived universe. In this framework, "place transcends materiality, even if not dissociated from it, because spatial memories eternalize in our memories" (MELLO, 2002, p. 63).

2.2   Spatialities and temporalities

In the face of a reality data or information about some event, two questions are inevitable: when did it occur? In what place? These questions are due to the fact that every event focuses on determined spatialities and specific temporalities (SANTOS, 2002). Thinking of place as a composite of spatialities and temporalities refers to the recognition of its dynamics generating and involving forms, functions, contents (ABREU, 2003) and meanings (TUAN, 2013) inserted in multiple levels of investigation.

For a long period of time, geography has privileged spatial analysis without paying due attention to time (SALGUEIRO, 2003). Today, however, the spatial and temporal dimensions must be valued (ABREU, 2003). The categories spatiality and temporality obviously lead us to inquire about space and time (GEIGER, 2003). However, spatiality goes beyond the rigidity associated with space as a mere stage of events, representing an advantage in the approach to spatialization of a dynamic phenomenon such as urbanization, for example. Similarly, we can speak of temporality by referring to the different periods (SALGUEIRO, 2003; FERNANDES, 2010). In geographical approaches it is not enough to unveil the multiple dimensions of contemporary spatiality. It is necessary to investigate - equally - the past temporalities responsible for the already configured inheritances. This is the path that the geographer must follow, if he wants a geography genuinely committed to spatiality (ABREU, 1997).

Space and time are basic categories of human existence (HARVEY, 1992), since they coalesce into different temporalities. In the approaches, it becomes indispensable to introduce temporality into space. In this reading, space and time are not reducible to each other. They are distinct but complementary terms (MASSEY, 2008).

The actions of nature and of the human being are inscribed in space (spatiality) and time (temporality). The phenomena derived from natural, social and existential influences spread, are located in spaces and places through their limits and dimensions. Through these actions in space and time, the different places emerge. Spatialities, as well as temporalities, emerge from the differentiation of the action of nature and the human being on the Earth's surface (CORRÊA, 2011).

In the words of Cosgrove (2004, p.110), "above all, a historical and contextual sensitivity on the part of the geographer is essential. We must resist the temptation to displace the landscape from its context of time and space".

The spatial and temporal dimensions of life have always been at the core of the humanistic perspective in geography. Both Lowenthal (1982; 1985; 1998) and Tuan (1980; 1983; 2011; 2013) - researchers pointed out by Holzer (1992, p. 489) as "the undisputed fathers of humanistic geography" - privileged the mentioned categories in the studies that founded the alluded current of geographical thought. In the opinion of Lowenthal (1982), time exerts a strong influence on our individual perspectives in relation to our particular universes. For the thinker, "each particular world has had a career in time, a history of its own" (p.138), in this direction, "every personal history results from a particular environment" (p.139). The relationship we nurture with our lived universe is built on our lived experience, and this demands time. Lowenthal concludes his mental elaboration by emphasizing that "all kinds of experiences (...), come together to form our individual picture of reality" (p.141). In the course of time, "as artists", we can create and/or organize spaces and places (LOWENTHAL, 1982, p.141).

Tuan (1980; 1983; 2011; 2013) also delegates indisputable relevance to the spatial and temporal dimensions, raising them to the level of the "framework" of humanism in geography (TUAN, 2011, p.8). In his conception,

Tuan points out that, "in the imagination it is easy to treat space" and "time separately". "In lived experience," however, "they are inextricably linked" (TUAN, 2011, p.8). "Movement demands time and occurs in space" (p. 15). "The longer we stay in a place, the better we know it and the more deeply it will become meaningful for us" (p. 17). For Tuan (2011, p. 18), space and time "are overlapping categories of human experience. If they are not considered together, the world of geographers will retain an air of unreality, abstaining from life as it is lived.

Time is implicit in all spaces and places. We all have a sense of space and time. "The ease with which we confuse the spatial and temporal categories is evident in language". "Often length is given in units of time" and "the passage of time" is described as length. "Time is still volume": the great moments of life (TUAN, 2013, p.147). "Space is historical". In this sense, "space and time have always been structured according to individual human feelings and needs" (TUAN, 2013, p. 152-153). "Space has temporal significance," both "in the poet's reflections" and "at the level of everyday personal experiences" (TUAN, 2013, p. 156). "Space and time coexist, intermingle and each is defined according to personal experience" (TUAN, 2013, p.161). "All that we are is due to the past. The present also has value, it is our experiential reality" (TUAN, 2013, p.239).

In his reflections in search of a geography that is not limited to scientific and academic   rigors, Wright (1947) indicates the need for geographers to be concerned with both spatiality and geography studies. The latter, understood as the deep involvement of individuals and social groups with their lived universe, this being the intimate spatiality of each human being. In this path, both spatiality (characteristics and dynamics of places) and geography (experiences, experiences, meanings and symbolisms) are linked to our existence. For this reason, it is worth repeating, one should not neglect the ambivalence and complementarity of these essential data of reality (WRIGHT, 1947; DARDEL, 2011).

2.3   Place and Symbolism

A symbol has the power to suggest a whole, transcends its condition as such and as an integral part is confused with the place in which it finds itself. In this regard, the symbolic charge of a temple or a stadium can be much wider, expressive than its original destination. In fact, the cross symbolizes Christianity, the crown the monarchy (TUAN, 2012) as well as the Brandenburg Gateway represents one of the greatest symbols of the German nation (FREITAS, 1999). The symbolism, understood as emblem or interpretation of the meaning of a certain symbolic element (symbol), is manifested in recent decades as a concept extremely important for humanistic and cultural research - studies related to the understanding of the subjective dimension of the place (TUAN, 1980; 2012; MELLO, 2000, 2003; FERNANDES, 2015). According to Cosgrove (2004):

> All landscapes have symbolic meanings because they are the product of man's appropriation and transformation of the environment. The symbolism is more easily apprehended in the most elaborate landscapes - the city, the park and the garden - and through the representation of the landscape in painting, poetry and other arts. But it can be read in rural landscapes and even in the most apparently non-humanized landscapes of the natural environment. The latter are often powerful symbols in themselves (COSGROVE, 2004, p.108).

To understand the expressions printed by a culture in its landscape, we need a knowledge of the language used: the symbols and their meaning in that culture. For the author, although the link is very tenuous between the symbol and what it represents, all landscapes are symbolic. By emphasizing that human scenarios are charged with symbolism, Cosgrove focuses on nature and the natural landscape as powerful symbols in themselves, based on the assumption that any human intervention in nature involves its transformation into culture. Although this transformation is not always visible, especially for a stranger, the natural object becomes a cultural object when it is given symbolic meaning (COSGROVE, 2004). Let us now observe Tuan's words (2012):

> Of the many and varied reasons to move to the suburbs, the search for a healthy environment and an informal lifestyle are among the oldest. We have repeatedly observed how the feeling for nature and rural life is encouraged by the pressures of urban life. The city environment is both seductive and irritating, beautiful and unpleasant. The rich have always been able to escape this by going out to rest in their country houses. In the western world the feeling for nature culminated with the romantic movement of the 18th and 19th centuries (...). The city symbolized corruption (...). The countryside symbolized life: life revealed in the fruits of the earth, in the green things that grow, in pure water and clean air, in the healthy human family (TUAN, 2012, pp. 324-325).

The symbology is not restricted to the centers of good will, affectivity, stripping or experience, because vast, strange and distant spaces are configured as symbols of rejection (TUAN, 1980; 2012; MELLO, 2003). Understanding symbolism as the framework of an idea - both negative and positive - of a certain symbolic element, Tuan (2012) proposes a counterpoint between the city and the countryside, suggesting that after the industrial revolution, the city - little by little - ceases to symbolize an ideal of life, giving to the countryside this condition, through a return to the feeling for nature. According to Tuan, by acquiring some of the values of the countryside, the suburb - understood as the frontier of metropolitan expansion - becomes an ideal, since it suggests a perfect lifestyle in which the best of rural and urban life is combined without its defects (TUAN, 2012). In this sense, whether for the American context above or within the Brazilian suburbs, notably the Carioca, the metropolitan peripheries now represent for their residents a symbol of good will (CORRÊA, 2000; FERNANDES, 2006; SOUZA, 2005).

Places and symbols acquire deep meaning through the emotional ties woven over the years. The place itself constitutes a symbol of affectivity, good will, satisfaction, happiness and harmony, on the one hand, but also a stage of struggles and daily life. The symbolic character of places establishes connections by decoding and translating its past and connecting it to the present (MELLO, 1990; 2003).

Still regarding the symbolic universe, let's consider the lucubrations of the geographer Doreen Massey (2008):

> And, thus, there is "place". In the context of a world that is certainly more and more interconnected, the notion of place (usually cited as "local place") has acquired a totemic resonance. Its symbolic value is incessantly mobilized in political arguments. For some, it is the sphere of everyday life, of real and valued practices, the geographical source of meaning, vital as a point of support, while "the global" weaves its webs, increasingly powerful and alienating. For others, "a refuge in place" represents the protection of drawbridges and the construction of walls against new invasions. Place, through this reading, is the place of denial.

In an attempt to translate the symbolic value of the place, Doreen Massey (2008, p.24-25) discusses its wide range of meanings. In her perspective, the place symbolizes - among other things - the sphere of everyday life, the geographical source of meaning, vital point of support, besides representing refuge and protection against the powerful and alienating webs of the global. Defending a new stimulus of spatiality, the author points out nature and the natural landscape as symbolic foundations for the recognition of place (MASSEY, 2008).

Entering this universe of meanings and values, Joel Bonnemaison (2002, p.109-111) stresses:

> A geosymbol can be defined as a place, an itinerary, an extension which, for religious, political or cultural reasons, in the eyes of certain people and ethnic groups assumes a symbolic dimension that strengthens them in their identity (...). Symbols gain greater strength and prominence when they are embodied in places. The cultural space is a geosymbolic space, loaded with affectivity and meaning.

The geosymbol - a concept worked on by Joel Bonnemaison (2002, p.109-111) - can be understood as a place-symbol, loaded with affectivity and meaning. Among the premises defended by geographers of the humanistic horizon are those related to the symbolic content of places (COSTA, 2008) and the mosaic of symbols that reside in it (MELLO, 2003, 2008). In this sense:

> The symbolic character of places reveals itself to the human being as something that precedes language and discursive reason, thus presenting certain aspects of reality, emphasizing the relationships between the symbolic and the place. These relations are mediated by symbols that can be a material reality and that join an idea, a value, a feeling. We understand, therefore, that symbolic mediations permeate personal attitudes towards places (COSTA, 2008, p 149).

In his discourse on the question of cultural heritage as a set of symbols that refers to the memory of the place, Costa (2008) alludes to the fact that "the symbolic of places leads us to the concept of the vernacular landscape where such a character is explicit in the set of representations, both of ancient landscapes and current ones, expressed through the knowledge and deeds of man" (p.151). For the author, certain elements of a natural or cultural order, when associated with the daily relationships of individuals or social groups, can define a set of symbols that express the memory of the place. In these circumstances, daily relationships and the consequent understanding of places and their symbols can make a space become a place, once affectively cut out. "In this context, the place becomes more interesting as a reference of identity and at the same time acquires a symbolic value" (COSTA, 2008, p. 155).

2.4   Polyvocality: the multiple interpretations of the lived universe

There is no ignoring the subtle and complex ties that bind the human being to his place. The symbolic and the other subjective connotations are present in the practice of interpreting and analyzing the spaces and places in their meanings. The human being is an informal geographer (LOWENTHAL, 1982) and the place his lived world in which relationships are mixed in a tangle of ties, experiences and clashes, where personal feelings, collective memories and symbols are present. From this point of view, we can understand the place as a reservoir of symbols to be interpreted, both by ordinary individuals and, consequently, by geographers. The place, as a humanized shelter, allows - with this - multiple readings concerning different manifestations that express as many interpretations as there are subjective and intersubjective meanings.

The places of human beings are composed of several layers of meanings. These multiple levels make the lived universe a symbolic place exposed to interpretation. From this perspective, "geography is everywhere" (COSGROVE, 2004, p. 96) and each individual, by thinking and seeking to understand his or her lived world, becomes an informal geographer (COSGROVE, 2004; LOWENTHAL, 1982), able to interpret his or her particular universe.

In Lowenthal's conception (1982, p. 105) "any person who examines the world around him is, in some way, a geographer". For the aforementioned thinker, the experts of this knowledge need to consider that there are other people decoding the world under different perspectives and that the vision of the lived universe should not be obtained only from one perspective. Each conception of the world is unique and there is no possibility of (LOWENTHAL, 1982; CORRÊA, 2007). In this rhythm, one of the reasons for such differentiations about what is lived is because all information is inspired and edited by the feeling of individuals for their place (LOWENTHAL, 1982). These are immensely rich accumulations or additions on individuals and social groups.

According to Geertz (2013), man is tied to webs of meaning woven by him. In order to decode them, scientific laws must give way to interpretative philosophies that translate a range of expressions inherent to all human activity, both material and symbolic (COSGROVE, 2003; GEERTZ, 2013).

Based on Hall (1997), Corrêa (2007, p.5) asserts that "symbols are open to different interpretations, each based on the experience, values, beliefs, myths and utopias of the social group that interprets them". The meanings attached to the symbols vary from person to person. This condition promotes polyvocality, that is, the different interpretations about the same symbol. Polyvocality becomes a kind of antidote to a single or unilateral imposed meaning (CORRÊA, 2007).

The revelation about the occurrence of multiple layers of meaning linked to human relations (GEERTZ, 2013), as well as their overlapping, has long been neglected (PANOFSKY, 2004), pointing to a pressing need: approaches that privilege the human element in its multiplicity, since the place is multidimensional.

2.5   The place in its multidimensionality

According to Relph (2012), the place has been a concern of philosophers since classical antiquity. Plato considered it as "the food of being, while others brought it closer to a geographical sense as the context in which beings are gathered together" (RELPH, 2012, p. 18). From the 17th century onwards, however, the Cartesian conception of space as a measurable dimension excluded the place of philosophy and sciences. Only in the 20th century did phenomenologists - "especially Husserl, Heidegger, Merleau-Ponty" (RELPH, 2012, p. 18) - identify the profound inadequacies of Cartesian logic. In the conception of these thinkers, positivism promotes the incision of philosophy by leaving out feelings, emotions, experiences of human beings. Positive science, by reducing spaces and places to a single dimension (material), promoted a deficient geography (RELPH, 2012). The aforementioned indifference to the subjective dimension of human existence is, according to Relph (2012), one of the possible motivations for the recent interest in the place. However, the above-mentioned geographer does not fail to point out that the concept in question has multiple dimensions. For Relph, besides bringing together the qualities, experiences and meanings of our experience, the place has other subjective aspects (spirit, meaning, rooting, interiority...) and also objectives (location, physiognomy, constructions...). The essence of the place presupposes the observance of its multidimensionality (RELPH, 2012). In her debate on the significant dimensions of place, Lívia de Oliveira (2012, p.12) points out that "every place acquires identity through its various spatial dimensions: location, direction, orientation, relationship, territory, spatiality and others". In the text in question, Professor Oliveira seeks a junction between the physical, experiential and symbolic dimensions in order to sustain the different spheres or layers of meaning "that fit the constitution of the place" (MARANDOLA JR, 2012, p. 15).

In their laborious reflections on place and subject, Berdoulay and Entrikin (2012) agree that the intense feeling of belonging, which creates a fusion between the individual and his lived world, can be promoted by measurable elements or by ephemeral existential relationships that make up our daily life. For the thinkers mentioned, by promoting changes in their place, human beings transform the Earth into their world. These changes, in turn, equally affect individuals and social groups in their lives of relationships. In this sense, the subject (human being) and the place (objective and subjective reality) form an inseparable whole (BERDOULAY AND ENTRIKIN, 2012). The place - lived universe, par excellence - goes beyond the geographical objectivity that ignores the subtleties responsible for its value aspects (GALLAIS, 2002; FERNANDES, 2010). What is lived also carries with it a dimension richer in subjective and existential aspects. It is a sphere of recognition and familiarity pertinent to everyday life in which its significance deepens its (geo)symbolic role (BONNEMAISON, 2002). In the words of Dardel (2011), geography, the work of man, displays a constructed space and possesses meaning. These constructions exalt man by intertwining his customs, habits, conduct, ideas and feelings in his surroundings and horizon. The different constructions differ in quality and meaning. Among the most important elements of the built space is man's habitat, his home, loaded with existential and symbolic values (SCHUTZ, 1979; TUAN, 2013; DARDEL, 2011). The place is the synthesis between man and the Earth in its subjectivity, values and meanings. This geography represents, according to Dardel (2011), "the nature of geographical reality". Despite its multiplicity, the place is unique. As subjects, human beings build places of belonging and identity and are, at the same time, shaped by such places. This exchange subject x object, between individuals and their places (inseparable elements), "builds obstacles to the postmodern and metropolitan tendency to see each place as the summary of all others" (BERDOULAY AND ENTRINKIN, 2012, p.112). The place has several dimensions (DARDEL, 2011), representing the basis of human existence and center of meaning, a postulation suggested by Entrikin (1980). In Bachelard's (2008) reflections, the place carries the essence of the notion of home and top (spelling) of our intimate being. In another current of thought, we find in Santos (2002) the following formulation for the place: obligatory depository of the event and irreplaceable theater of human passions. In this context, it becomes the basis for the reproduction of life and the product of human relations according to Carlos (1996). Besides this, the place in Armando Corrêa da Silva's words gains the property of a spatial dimension in its entirety and totality of what could be supposed to be reality (SILVA, 1988). According to the canons of humanistic perspective, space endowed with value and meaning emerges at the level of place or home (TUAN, 2013). Through the reading of its multiple definitions, place can be considered a lived universe, not only in its concreteness and objectivity, but with all the biases of human imagination (BACHELARD, 2008).

In its multidimensionality, the place is a hybrid entity composed by materiality, symbolism, spatialities, temporalities and the existential way of experiencing it. To reconcile and decode this multiplicity that represents the nature of geography (DARDEL, 2011) was the task I proposed to develop in this book.

FINAL WORDS

Throughout this book, we have seen that several social scientists - especially after the 1970's - have been committed to criticizing the reductionist vision of man, postulated by positive science, a tendency that favored humanistic geographers the interpretation of feeling and understanding of the relations between men and their world. This current of geographical thought is also opposed to positivism, in an attempt to overcome its reductionism. Some philosophers like Wilhelm Dilthey (1833-1911), however, in the passage between the 19th and 20th centuries, were already critical of the scientific, naturalistic and positivist tendency that guided the human sciences of the time. For the aforementioned scientist, a precursor of hermeneutics, the facts concerning the spirit or the soul are not similar to natural processes, since they refer to the human world of meaning and value. Thus, the niche of valuations and other feelings coming from subjectivity and intersubjectivity should not be removed from the historical context of individuals and social groups, since it is not possible to formulate objective laws about it, but rather to seek their understanding and interpretation (ARANHA, 1996; ARANHA; MARTINS, 1992; JAPIASSÚ, 1975).

In this journey, positivism would be responsible for a fable, since through this inclination, science becomes a myth, being considered the only adequate form of knowledge to the detriment of other possible approaches to reality. It is, therefore, a deformed perspective of knowledge because, by admitting that true knowledge refers only to what can be proved, experienced and, therefore, objective, it incurs a reductionism that limits the approaches referring to the human being - who, by thinking, feeling, dreaming and philosophizing - suggests that his subjectivity is the focus of scientific research. In fact, it was from the pretended objectivity proclaimed by the positivist ideology that derived the "myth of scientific neutrality," according to which scientific research would be on the margins of social, cultural or political influence, occupying the scientist with the description of phenomena, without interfering in studies.

Phenomenology, as we know, was one of the first philosophies to oppose reductionism and the scientificism of so-called positive sciences. While positivism requires an increasingly neutral scientific knowledge, devoid of subjectivity and, therefore, distant from man, phenomenology proposes the resumption of the humanization of science, with a new relationship between subject-object and man-world, considered inseparable halves. As a donor of meaning and source of meaning to the world, the consciousness is not restricted to mere intellectual knowledge, but is the generator of intentionalities that are not only cognitive but also affective and practical. The look upon the world is the act by which man experiences his lived universe, imagining, judging, loving, fearing, dreaming. In this sense, the universe I experience is a world for me, hence the importance of meaning, of the network of meanings that involves what is sensory, mental or psychologically captured or understood (ARANHA, 1996; ARANHA; MARTINS, 1992; JAPIASSÚ, 1975).

The mythical image of the scientist ignores that he is part of and depends on a real structure of the world around him. This supposed neutrality ignores the persuasive power of the experience, as if the researcher could be the holder of a single truth that, once formulated in its coherence, would be free from questioning; as if he could keep forever the image of an individual immune to the incoherence of passions (JAPIASSÚ, 1975). Due to this positivist framework which, either consciously or unconsciously, persists in permeating most of the approaches related to the humanities, methodological concerns such as "the use of the same verbal person from the beginning to the end of the text" and - above all - "the necessary extra care not to incur in the fateful error of partiality", unduly interfering in the research which, despite being of his authorship, should not manifest his particular point of view.

Like most of those who focus on approaches to human phenomena, I have always strived to follow the main models and rules set out above. I remember that during the time I was writing my specialization monograph, in a conversation with a classmate, I asked the following question: how can I be impartial, addressing a phenomenon that also influences me, in a world with which I also share? In an attempt to console myself, my friend then answered such a question by saying that "nothing can be more humanistic than a study done by a researcher in his own place. On the occasion of the period described, I was still unaware of the existence of the letter of jail to the kind of transgression that worried me so much. This freedom is proclaimed through the philosophies of the meaning that guide humanistic geography, which - contrary to positivist rules, laws, models, certainties, precision - defends that the researcher, involved in the lived universe of people, can either use different verbal people (me or us, for example), or compromise the pretended impartiality of the research, defended by positivist and neo-positivist paradigms.

The widespread "scientific neutrality" has become a myth due to the discernment that all scientific research defends a certain point of view, both of the researcher and of the group he represents, and is therefore biased.

In order not to end this book amidst methodological concerns and justifications, I would like to evoke the sacred character of the sense of place, basing myself on the entrenched religious perspective of Mircea Eliade (2007) in relation to the mythical description of the symbolism of the center. The allusion to this sacred approach represents a last attempt to express - through words - the value, meaning, importance and relevance of the place for the person who experiences it and makes it his central axis to where everything converges. The aforementioned philosopher, in dedicating himself to the history of religions, travels through some oriental traditions until situating Paradise, where man (Adam) was created by God from clay, in the center of the world - or in the navel of the Earth (ELIADE, 2007; 2012). This insight, brought by Eliade, takes us back to a kind of reverie in an attempt to find out the reason for the love and veneration of the human being for his lived universe.

What would have been the greatest punishment imposed on man for his disobedience to the Divine order? Some would say it was the exhaustion imposed by work that from then on became tiring. Most would certainly claim to be death, which came into force after sin, the "capital punishment" imposed on man. Physical death, however, was not imposed by the Creator, surviving to man as the greatest consequence of his sin. Once death was discarded, wouldn't the expulsion from Paradise - the place par excellence - have been the great punishment imposed on man by the Creator?

REFERENCES

ABBAGNANO, Nicola. *Dictionary of Philosophy*. São Paulo: Martins Fontes, 2007. 1210 p.

ABREU, Maurício de Almeida. *Urban evolution in Rio de Janeiro*. Rio de Janeiro: IPLANRIO, 1987. 147 p.

______. The City, the Mountain and the Forest. In: ABREU, Maurício de Almeida. *Nature and Society in Rio de Janeiro*. Rio de Janeiro: Biblioteca Carioca, 1992. p. 55-103.

______. The Appropriation of Territory in Colonial Brazil. In: CASTRO, Iná Elias de; GOMES, Paulo César da Costa; CORRÊA, Roberto Lobato (Org). *Geographical Explorations:* Routes at the End of the Century. Rio de Janeiro: Bertrand Brasil, 1997. p.197-245.

______. About the Memory of Cities. *Revista da Faculdade de Letras - Geografia I série*, Porto, v.14, p. 77-97, 1998.

______. Cities: Spatial and Temporalities. In: CARLOS, Ana Fani Alessandri; LEMOS, Amália Inês Geraiges (Orgs). *Urban Dilemmas:* new approaches on the city. São Paulo: Context, 2003. 430 p.

______. *Urban evolution of Rio de Janeiro*. Rio de Janeiro: IPP, 2008. 155 p.

ABREU, Regina. Gum I mix with bananas? About the relationship between theory and research in social memory. In: GONDAR, Jô; DODEBEI, Vera (Orgs). What *is social memory?* Rio de Janeiro: Contra Capa Livraria, 2011. 160 p.

ALMEIDA, Maria Geralda de. The songs and enchantments of a country geography of Patativa do Assaré. In: MARANDOLA JR, Eduardo; GRATÃO, Lúcia Helena Batista (Orgs). *Geography and Literature*: essays on geography, poetics and imagination. Londrina: Eduel, 2010. 354 p.

ANTONELLO, Ideni Terezinha. The Amazonian territorialities shine in the literary narrative of Peregrino Júnior. In: MARANDOLA JR, Eduardo; GRATÃO, Lúcia Helena Batista (Orgs). *Geography and Literature*: essays on geography, poetics and imagination. Londrina: Eduel, 2010. 354 p.

ARANHA, Maria Lúcia de Arruda. *History of* Education. São Paulo: Modern, 1996. 255 p.

ARANHA, Maria Lúcia de Arruda; MARTINS, Maria Helena Pires. *Themes of Philosophy*. São Paulo: Editora Moderna, 1992. 232 p.

ARAÚJO, Dorothy Sue Dunn de. The Vegetation of the Guaratiba-Sepetiba baixada. In: KNEIP, Maria Lina et al. *Guaratiba-Rio de Janeiro Prehistoric Collectors and Fishermen*. Rio de Janeiro: UFRJ; Niterói: EDUFF, 1987. p. 47-72.

ASSIS, Lenilton Francisco de. Second Residence Tourism: The Spatial Expression of the Phenomenon and the Possibilities of Geographic Analysis. *Território magazine*, Rio de Janeiro: set/out, p. 107-122, 2003.
ATLAS of the nature conservation units of the state of Rio de Janeiro, 1990. Irregular pagination.

BACHELARD, Gaston. *The poetics of space*. São Paulo: Martins Fontes, 2008. 242 p.

BANDEIRA, Lucia Batista. Mallet - *A Neighborhood Elected and Demarcated Effectively*. 1999. Monograph (Specialization in Territorial Policies in the State of Rio de Janeiro) - Instituto de Geografia, Universidade de Estado do Rio de Janeiro, Rio de Janeiro, 1999.

BASTOS, Ana Regina Vasconcelos Ribeiro. Space and Literature: some theoretical reflections. *Espaço e Cultura magazine*, Rio de Janeiro, n.5, 1998.

BELL, Daniel. *The Advent of the Post-industrial Society*. São Paulo: Ed. Cultrix, 1977.

BERDOULAY, Vicent; ENTRIKIN, John Nicholas. Place and Subject: theoretical perspectives. In: MARANDOLA JR, Eduardo et al. *What is the space of the place?* Geography, epistemology, phenomenology. São Paulo: Perspective, 2012. 307 p.

BERTA, Ruben. Guaratiba: Urban Structuring Plan foresees buildings of 4 floors. *O Globo Newspaper,* Rio de Janeiro, 28 Apr. 2012.

BONNEMAISON, Joel. Trip around the Territory. In: CORRÊA, Roberto Lobato; ROSENDAHL, Zeny (Org). Cultural Geography: a Century (3). Rio de Janeiro: EdUERJ, 2002. p. 83-131.

BORBA, Francisco da Silva. *UNESP Dictionary of contemporary Portuguese.* Curitiba: Piá, 2011. 1488 p.

BOSSÉ. Mathias Le. Identity issues in cultural geography - Some contemporary conceptions. In: CORRÊA, Roberto Lobato; ROSENDAHL, Zeny. *Cultural Geography* - An Anthology. V. 2. Rio de Janeiro: EdUERJ, 2013. 292 p.

BOSI, Ecléa. *Memory & Society* - Remembrance of old people. Companhia das Letras: São Paulo, 2003. 488 p.

BUTTERFIELD, Roger. "Henry Ford, the Wayside Inn, and the Problem of 'History is Bunk'". *Massachusetts Historical Society Preccedings*, Massachusetts, v. 77, p. 53-66,1965.

BUTTIMER, Anne. Seizing the Dynamism of the Living World. In: CHRISTOFOLETTI, Antônio (Org). *Perspectives of Geography.* São Paulo: DIFEL, 1982. p. 165-193.

BUTTIMER, Anne. Home, Horizon of Reach and the Meaning of Place. *Geography Magazine*, Niterói, v.5, n.1, p. 4-19, summer 2015.
CALS, Soraia. *Roberto Burle Marx*: a Photobiography. Rio de Janeiro: Art Scholarship, 1995. Irregular pagination.

CANCLINI, N. G. *The Imagined Globalization*. Buenos Aires: Paidós Publishing House, 1999.

CAPEL, Horácio. *The Cosmopolis and The City*. Barcelona: Del Serbal Publishing House, 2003.
CARLOS, Ana Fani Alessandri. *The Place in/out of the World*. São Paulo: Hucitec, 1996.

CARVALHO, Ronaldo Cerqueira de. *Rio de Janeiro:* A city connected by tunnels - panorama until the end of the sixties. 2002. Monograph (Specialization in Geography) - Instituto de Geografia, Universidade do Estado do Rio de Janeiro, Rio de Janeiro, 2002.

CARVALHO, Ronaldo Cerqueira de. Rio de Janeiro - a city connected by tunnels. Rio de Janeiro: Municipal Department of Urbanism; Instituto Municipal de Urbanismo Pereira Passos, 2004. 57 p.

CASSIRER, Ernest. *The Philosophy of Symbolic Forms*: II - the mythical thought. São Paulo: Martins Fontes, 2004. 432 p.

______. *The Philosophy of Symbolic Forms*: III - phenomenology of knowledge. São Paulo: Martins Fontes, 2011. 818 p.

CASTELLS, Manuel. *The Network Society*. São Paulo: Peace and Earth, 2002.

CASTRO, Augustus Caesar de. *Guaratiba*: Yesterday and Today. Graduation Work (Graduation in History) - Graduation Course in History, Unified Educational Foundation Campograndense (FEUC), Rio de Janeiro, 2002.

CHRISTOFOLETTI, Antônio. The Perspectives of Geographical Studies. In: CHRISTOFOLETTI, Antônio. Perspectives *of Geography*. São Paulo: DIFEL, 1982. p. 11-36.

CLAVAL, Paul. The Role of the New Cultural Geography in the Understanding of Human Action. In: CORRÊA, Roberto Lobato;

ROSENDAHL, Zeny (Org). *Matrices of Cultural Geography*. Rio de Janeiro: EdUERJ, 2001. p. 35-86.

______. *Cultural Geography*. Florianópolis: Publisher of UFSC, 2007. 453 p.

CORRÊA, Magalhães. The Sertão Carioca. *Magazine of the Brazilian Historical and Geographic Institute,* Rio de Janeiro, v. 167, 478p., 1936.

CORRÊA, Roberto Lobato. The Environment and the Metropolis. In: ABREU, Maurício de Almeida. *Nature and Society in Rio de Janeiro*. Rio de Janeiro: Biblioteca Carioca, 1992. p. 27-36.

______. *The Urban Space*. 4.ed. São Paulo: Attica, 2000. 94 p.

CORRÊA, Roberto Lobato . Space: A Key Concept of Geography. In: CASTRO, Iná Elias de; GOMES, Paulo César da Costa; CORRÊA, Roberto Lobato (Org). *Geography*: concepts and themes. 4.ed. Rio de Janeiro: Bertrand Brasil, 2002. p. 15-47.

______. Cultural Geography and the Urban. In: CORRÊA, Roberto Lobato; ROSENDAHL, Zeny (Org). *Introduction to Cultural Geography*. Rio de Janeiro: Bertrand Brasil, 2003. p. 167-186.

______. About cultural geography. In: NEPEC texts (volume 3). Rio de Janeiro: UERJ, 2007.

CORRÊA, Roberto Lobato. The spatiality of culture. In: CD culture. Rio de Janeiro: UERJ, 2011.

CORRÊA, Roberto Lobato; ROSENDAHL, Zeny. Cultural Geography: presenting an anthology. In: CORRÊA, Roberto Lobato; ROSENDAHL, Zeny (Orgs). *Cultural Geography:* an anthology (1). Rio de Janeiro: EdUERJ, 2012. 344 p.

COSGROVE, Denis. Towards a radical cultural geography: problems of theory. In: CORRÊA, Roberto Lobato; ROSENDAHL, Zeny (Org). *Introduction to cultural geography*. Rio de Janeiro: Bertrand Brasil, 2003. 224 p.

______. Geography Is Everywhere: Culture and Symbolism in Human Landscapes. In: CORRÊA, Roberto Lobato; ROSENDAHL, Zeny (Org). *Landscape, Time and Culture*. Rio de Janeiro: EdUERJ, 2004. p. 92-123.

COSTA, Mariana Timóteo da. Two steps away from paradise. *O Globo Magazine*, Rio de Janeiro, 15 Dec. 2013.

COSTA, Otávio. Memory and Landscape: in search of the symbolic of places. *Espaço e Cultura magazine,* Rio de Janeiro, commemorative edition 1993-2008, p. 149-156, 2008.

DARDEL, Eric. *Man and the Earth*: nature of geographical reality. São Paulo: Perspective, 2011. 159 p.

DIAS, Alice Ferreira Rodrigues. Guaratiba Island: Green Landscape for Who? *Espaço Aberto Magazine*, Rio de Janeiro, v.1, n.1, p. 89-108, 2011.
DREW, David. *Interactive Processes Man-Mean Environment*. Rio de Janeiro: Bertrand Brasil, 2002. 224 p.

ELIADE, Mircea. *La Nostalgie des Origines*. Paris: Folio-Essais; Galimard, 1971.

______. *Myth of eternal return*. São Paulo: Mercuryo, 2007. 175 p.

ELIADE, Mircea. *Images and symbols*: essay on magic-religious symbolism. São Paulo: Martins Fontes, 2012. 178 p.

ELIAS, Norbert & SCOTSON, John. *The Established and the Outsiders*: Sociology of Power Relations from a Small Community. Rio de Janeiro: Jorge Zahar, 2000 224 p.

ENTRIKIN, John Nicholas. Contemporary Humanism in Geography. *Boletim de Geografia Teorética*, São Paulo, v. 10, 1980.

FARAH, Fábio. José Humberto Resende - The Shroud Doctor. *Isto é Gente Magazine*, São Paulo, June 2, 2003. Available at: http://www.terra.com.br/istoegente   / 200/reportagens/jose_humberto_resende_sudario.htm>. Access on: 20 Nov. 2015.

FERNANDES, Marcio Luis . *Guaratiba Island*: From Space to Place. 2003. 44f. Monography (Degree in Geography) - Moacyr Sreder Bastos University Center, Rio de Janeiro, 2003.

FERNANDES, Marcio Luis. *Guaratiba Island* in its natural attributes. Rio de Janeiro, 2005. 41f. Available at: <http://www.webartigos.com/artigos/ilha-de-guaratiba-em-seus-atributos-naturais/136345/>. Access on: 19 Oct. 2015.

______. The *Valuation of "Space" producing the valuation of "Place":*" The case of Ilha de Guaratiba - R.J. 2006. 56 f. Monography (Specialization in Geography) - Instituto de Geografia, Universidade do Estado do Rio de Janeiro, Rio de Janeiro, 2006.

______. *Guaratiba Island and its landscapes*. Rio de Janeiro, 2008. 32 f. Available at: <http://www.webartigos.com/artigos/ilha-de-guaratiba-e-suas-paisagens/ 136627/>. Access on: 19 Oct. 2015.

______. For a Necessary Change of Values: a proposal for the production of a (urban) space that favors the use and not the exchange. In: NATIONAL SYMPOSIUM RURAL AND URBAN IN BRAZIL, 2., 2009, Rio de Janeiro. *Annals...* Rio de Janeiro: UERJ, 2009.

______. *Decoding past and present geographies of Ilha de Guaratiba*. 2010. 99f. Dissertation (Masters in Geography) - Instituto de Geografia, Universidade do Estado do Rio de Janeiro, Rio de Janeiro, 2010.

______. *Discovering the urbanization march in Ilha de Guaratiba from the experiences lived by its residents*. Rio de Janeiro, 2011. 25 f. Available at: http://www.webartigos.om/artigos/descortinando-a-marcha-urbanizadora - in Guaratiba Island - for life-experiences-by-your-owners/136505/>. Access on Oct. 19, 2015.

FERNANDES, Marcio Luis. The Identity Character of Toponymy. In: CONGRESSO INTERNACIONAL DO NÚCLEO DE ESTUDO DAS AMÉRICAS, 3., 2012. Rio de Janeiro. *Annals...* Rio de Janeiro: UERJ, 2012.

______. Globalization and urbanization of the world. Rio de Janeiro, 2013. Available at: <http://www.webartigos.com/artigos/globalizacao-e-urbanizacao-do-mundo/136480/>. Access on Oct. 19, 2015.

______. Another horizon in search of the humanization of geography. *Geography Magazine*, Niterói, v.4, n.1, p. 78-87, summer 2014.

______. West Side Race. *The Day*, Rio de Janeiro, Nov. 26, 2014 b. Opinion Column.

______. Discovering the symbolic universe of a place. Perspectiva Geográfica magazine, Cascavel, v.9, n.11, 2015.

FERREIRA, Antônia Maria M; OLIVEIRA, Marli Vieira de. Contribution to the Archaeological Study of the Quaternary Superior of the Baixada de Guaratiba-Sepetiba. In: Kneip, Maria Lina et al. *Collectors and Prehistoric Fishermen of Guaratiba-Rio de Janeiro*. Rio de Janeiro: UFRJ; Niterói: EDUFF, 1987. p. 29-45.

FIREY, Walter. Sentiments and symbolism as ecological variables. *American Sociological Review*, v.10, n.2, p. 140-148, 1945. (Annual Meeting Papers).

FIREY, Walter. Feelings and symbolism as ecological variables. In: ROSENDAHL, Zeny; CORRÊA, Roberto Lobato. *Cultural geography*: an anthology, volume II. Rio de Janeiro: EdUERJ; 2013. p. 21-34.

FREITAS, Inês Aguiar de; PERES, Waldir Rugero; RAHY, Ione Solomão. The Window of Hitler. *GeoUERJ - Geography Department Magazine*. Rio de Janeiro n. 6, p. 29-36, 1999.

FRÉMONT, Armand. History of a Research. In: Frémont et ali. *Espace Vécu et Civilisations.* Paris: CNRS, 1982. 106 p.

FRIDMAN, Fania. *Owners of Rio in the name of the king*: a land history of the city of Rio de Janeiro. Rio de Janeiro: Jorge Zahar Editor (Garamond Publishing House), 1999. 304 p.

GALLAIS, Jean. Some Aspects of Living Space in the Civilizations of the Tropical World. In: CORRÊA, Roberto Lobato; ROSENDAHL, Zeny (Org). *Cultural Geography*: one century (3). Rio de Janeiro: EdUERJ, p. 63-81, 2002.

GALVÃO, Maria do Carmo Corrêa. Focus on environmental issues in Rio de Janeiro. In: ABREU, Maurício de Almeida. *Nature and Society in Rio de Janeiro.* Rio de Janeiro: Biblioteca Carioca, 1992. p. 13-26.

GEIGER, Pedro Pinchas. Contribution to the debate on urban spatialities and temporalities. In: CARLOS, Ana Fani Alessandri; LEMOS, Amália Inês Geraiges (Orgs). *Urban Dilemmas*: new approaches on the city. São Paulo: Context, 2003. 430 p.

GEERTZ, Clifford. *The interpretation of cultures*. Rio de Janeiro: LTC, 2013. 213 p.

GOMES, Paulo César da Costa. *Geography and Modernity*. Rio de Janeiro: Bertrand Brasil, 2007. 366 p.
GONDAR, Jô; DODEBEI, Vera (Orgs). *What is social memory?* Rio de Janeiro: Bookstore Cover, 2011. 160 p.

WAR, Antônio Teixeira. *Geological and Geomorphological Dictionary*. 8. ed. Rio de Janeiro: IBGE, 1993. 446 p.

HAESBAERT, Rogério. *The Myth of Desterritorialization*: From the "End of Territories" to Multiterritoriality. Rio de Janeiro: Bertrand Brasil, 2004. 400 p.

______. From deterritorialization to multiterritoriality. In: LATIN AMERICA GEOGRAPHIC MEETING - EGAL, 10., 2005, São Paulo. *Anais...* São Paulo: USP, 2005.

HALL, Stuart. *Representations:* cultural representations and signifying practices. London: Routledge publications, 1997.

HALBWACHS, Mauritius. *La Mémoire Collective.* Presses Universitaires de France: Paris, 1968.

______. *The Collective Memory.* São Paulo: Centauro, 2013. 224 p.

HALLEY, Bruno Maia. The neighborhood and the plots of the place. Geography Magazine, Niterói, v.4, n.1, p. 43 - 57, Summer 2014.

HERDY, Jose Maria. Guaratiba Island has everything to become an efficient neighborhood.In: PORTAL Guaratiba, Rio de Janeiro, 01 jun. 2013. Available at: http://www.portalguaratiba .com.br/2013/news/010601_guaratiba_island_tem_tudo_para_se_tornar_um _bairro_efficient.html>. Access in: 20 nov. 2015.
HARVEY, David. *Post-Modern Condition.* São Paulo: Loyola Editions, 1989.

______. *Postmodern Condition*: a research on the origins of cultural change. São Paulo: Loyola Editions, 1992. 349 p.

______. *The Capitalist Production of Space.* São Paulo: Annablume, 2005.

HIGHET, Gilbert. *The Classical Tradition*: Greek and Roman Influences on Western Literature. Oxford: Clarendon, 1949.

HOLZER, Werther. *Humanistic geography:* its trajectory from 1950 to 1990 (Master's Dissertation). Rio de Janeiro: PPGEO/UFRJ, 1992. 550 f.

______. The Phenomenological Geography of Eric Dardel. In: CORRÊA, Roberto Lobato; ROSENDAHL, Zeny (Org). *Matrices of cultural geography*. Rio de Janeiro: EdUERJ, 2001. p. 103-122.

______. The Humanist Geography: a review. *Espaço e Cultura magazine,* Rio de Janeiro, commemorative edition 1993-2008, p. 137-147, 2008.

JANOT, Luiz Fernando. On the way to Guaratiba. *O Globo Newspaper,* Rio de Janeiro, 26 Oct. 2013. Opinion column.

JAPIASSÚ, Hilton; MARCONDES, Danilo. *Basic philosophy dictionary*. Rio de Janeiro: Zahar, 2006. 309 p.

JOHNSTON, Ronald John. Introduction to the international study of the history of geography. In: JOHNSTON, Ronald John; CLAVAL, Paul (Orgs). *Current Geography: geographers and trends*. Barcelona: Ariel, 1986 a. p. 13-25.

JOHNSTON, Ronald John. *Geography and Geographers*: the Anglo-American human geography since 1945. São Paulo: Difel, 1986. 359 p.

KNEIP, Maria Lina et AL (Org). *Prehistoric Collectors and Fishermen of Guaratiba-Rio de Janeiro*. Rio de Janeiro: UFRJ; Niterói: EDUFF, 1987. 257 p.

KORYTOWSKI, Ivo. Literature and Rio de Janeiro. Rio de Janeiro, [20--]. Available at: < http://literaturaeriodejaneiro.blogspot.com.br/>. Access on: 20 nov. 2015.

LAGO, Mário. *In the Roller of Time*. José Olympio: Rio de Janeiro, 1976. 350 p.

LEFEBVRE, Henri. The *Presence and The Absence*. Contribution to the theory of representations. Mexico: FCE, 1983. Paginação irregular.

______. *The Urban Revolution*. Belo Horizonte: UFMG Publishing House, 2008.

LESSA, Carlos. *O Rio de Todos os Brasis*: Uma Reflexão em Busca de Auto-Estima. 2. ed. Rio de Janeiro: Record, 2001. 478 p.

LOWENTHAL, David. Geography, Experience and Imagination: Towards a geographical epistemology. In: CHRITOFOLETTI, Antônio. *Perspectives of Geography*. São Paulo: DIFEL, 1982. p. 103-141.

______. *The Past Is a Foreign Country*. Cambridge: Cambridge University Press, 1985.

______. How We Know the Past. *Projeto História - Journal of the program of postgraduate studies in history*, São Paulo, v. 17, p. 63-201, 1998.

MARANDOLA JR, Eduardo. About Ontologies. In: MARANDOLA JR, Eduardo et al. *What is the space of the place?* Geography, epistemology, phenomenology. São Paulo: Perspective, 2012. 307 p.

MASCARENHAS, Gilmar. *The Place of the Free Trade Fair in the Great Capitalist City: Conflict, Change and Persistence (Rio de Janeiro: 1964-1989)*. 1991. 220 f. Dissertation (Masters in Geography) - Federal University of Rio de Janeiro, Rio de Janeiro, 1991.
MASSEY, Doren. *Through Space*: A New Space Policy. Rio de Janeiro. Bertrand Brasil, 2008. 312 p.

MELLO, João Baptista Ferreira de . Humanistic Geography: The Perspective of the Experienced and a Radical Critique of Positivism. *Revista Brasileira de Geografia*, Rio de Janeiro, p. 91-115, 1990.

MELLO, João Baptista Ferreira de. *The Rio de Janeiro of Composers of Brazilian Popular Music - 1928/1991 - an introduction to humanistic* geography. 1991. 300 f. Dissertation (Masters in Geography) - Federal University of Rio de Janeiro. Rio de Janeiro, 1991.

______. The Humanization of Nature - an odyssey for the (re)conquest of paradise. In: SILVA, S. T.; VIANA, O. M. *Geography and Environmental Issue*. Rio de Janeiro: IBGE, 31-40, 1993.

______. In Defense of Individuals in Geographical Studies. In: NATIONAL GEOGRAPHICAL PENSION HISTORY MEETING 1. 1999, Rio Claro. Anais...Rio Claro: UNESP, 1999. p. 113-118.

______. *From the Spaces of Darkness to the Places of Extreme Light* - The Universe of the Star Marlene as and document for the construction of geographical concepts. 2000. Thesis (PhD in Geography) - Federal University of Rio de Janeiro, Rio de Janeiro, 2000.

______. Discovering and (Re)Thinking Space Categories Based on the Work of Yi-Fu Tuan. In: CORRÊA, Roberto Lobato; ROSENDAHL, Zeny (Org). *Matrices of cultural geography*. Rio de Janeiro: EdUERJ, 2001. p. 87-101.

______. The Restoration of the Places of the Past. GeoUERJ Magazine, Rio de Janeiro, n.12. p. 63-69, 2002.

______. Symbols of Places, Spaces and "Places". *Space and Culture*. Rio de Janeiro, v. 16, p. 64-72, 2003.

MELLO, João Baptista Ferreira de. In the Pulsar of the Marvelous City of São Sebastião do Rio de Janeiro. In: SOCIEDAD LATINOAMERICANA DE ESTUDIOS SOBRE AMERICA LATINA Y EL CARIBE - SOLAR, 9., 2004, Rio de Janeiro. *Annals...* Rio de Janeiro: [S.n.], 2004.

______. Values in Geography and the Dynamism of the World Lived in Anne Buttimer's Work. *Space and Culture*. Rio de Janeiro, v. 19-20, p. 33-40, 2005.

______. The Drums and Arrows of São Sebastião do Rio de Janeiro. *Imaginary and Art Magazine,* São Paulo, n.15 p. 37-67, 2007.

______. The River of Official and Vernacular Symbols. In: CORRÊA, Roberto Lobato; ROSENDAHL, Zeny (Org). *Space and culture*: thematic plurality. Rio de Janeiro: EdUERJ, 2008. p. 173-186.

MENDILOW, Adam Abraham. *Time and the novel*. London: John Spencer/Badger, 1960.

MENEZES, Luiz Fernando et al (Org). *Natural History of Marambaia*. Rio de Janeiro: EDUR, 2005.

MENEZES, Luiz Fernando; ARAÚJO, Dorothy Sue Dunn de; GOES, Maria Hilde de Barros. Marambaia: the last preserved carioca sandbank. *Ciência Hoje magazine*, v. 23, n.136, p. 28-37, mar. 1988.

MONBEIG, Pierre. *New studies of Brazilian human geography*. São Paulo: European diffusion of the book, 1957.

MORAES, Antônio Carlos Robert de. *Geography: a* short critical history. São Paulo: Annablume, 2007. 152 p.

MOREIRA, Ruy. *The Brazilian geographic thought*: the matrices of renewal. São Paulo: Context, 2009. 172 p.

NOGUÉ Y FONT, J. El Paisaje existential of five groups of environmental experience. Methodological essay. In: BALLESTEROS, A. *Geography and Humanism*. Barcelona: Oikos-tau, 1992. p. 87-96.

NORA, Pierre. *Between memory and history:* the problem of places. São Paulo: PUC-SP. 1993. p. 7-28. (History Project, n. 10).

OLIVEIRA, Livia de. The sense of place. In: MARANDOLA JR, Eduardo et al. *What is the space of the place?* Geography, epistemology, phenomenology. São Paulo: Perspective, 2012. 307 p.

PALMER, Richard E. *Hermeneutica*. São Paulo: Martins Fontes, 1970. 284 p. (Editions 70).

PANOFSKY, Erwin. *Meaning of visual arts*. São Paulo: Perspective, 2004.

PARK, Robert Ezra. The City: suggestions for the investigation of human behavior in the urban environment. In: OLD, Otávio Guilherme. *The Urban Phenomenon*. Rio de Janeiro: Zahar, 1976. p. 26-67.

PICKLES, Jonh. *Geography and humanism*. Norwich: Geo Books, 1985. 64 p.
PINTO, Rivadavia. Guaratiba: A Pride of 407 Years. *Reason: the positive newspaper*. Rio de Janeiro, unpublished, Nov. 1986.

POCOCK, Douglas. Place and the novelist. *Transactions of the Institute of British Geographers*: New Series, v. 6, p. 87-98, 1981.
PORTAL Guaratiba. Rio de Janeiro, [20--]. Available at: <http://www.portalguaratiba.com.br>. Access in: 20 nov. 2015.

DIGITAL PROCESSING: Geotechnologies and free software. [S.l.], [20--]. Available at: <http://www.processamentodigital.com.br>. Access on: 20 nov. 2015.

REDONDO, Andrea Albuquerque. The city grows to Guaratiba. *Urbe Carioca*. Rio de Janeiro, 25 Sep. 2012. Available at:< http:// urbe carioca.blogspot.com.br>.
Access on: 20 Nov. 2015.

RELPH, Edward. An inquiry into the relations between phenomenology and geography. *Canadian Geographer.*, v. 14, n.3, p. 193-201, 1970.

______. *Place and Placelessness*. London: Pion, 1976. 156 p.

______. Reflections on emergency, aspects and essences of place. In: MARANDOLA JR, Eduardo et al. *What is the space of the place?* Geography, epistemology, phenomenology. São Paulo: Perspective, 2012. 307 p.

RHEINGANTZ, Carlos. *First Families of Rio de Janeiro (XVI and XVII centuries).* V. 1 Brazilian Bookstore Publisher: Rio de Janeiro, 1965. 556 p.
RIBEIRO, Luiz César de Queiroz. *From tenements to gated communities*: the production forms of housing in the city of Rio de Janeiro. Rio de Janeiro: Civilização Brasileira: IPPUR, UFRJ: FASE, 1997. 352 p.

RIBEIRO, Miguel Ângelo; COELHO, Maria do Socorro Alves. The importance of the second habitation phenomenon and its implications with the leisure-version activity: the example of the State of Rio de Janeiro. In: ENCONTRO NACIONAL DA ANPEGE, 9., 2007, Niterói. *Annals...* Niterói: ANPEGE, 2007. 1 CD-ROM.

RIO 2016. Rio de Janeiro, [20--]. Available at: <http://www.rio2016.com>. Access on: 20 nov. 2015.

ROSENDAHL, Zeny. *Space and Religion*: A geographical approach. Rio de Janeiro: EdUERJ: 2002. 90 p.

ROSENDAHL, Zeny. Space, Culture and Religion: Dimensions of Analysis. In: CORRÊA, Roberto Lobato; ROSENDAHL, Zeny (Org). *Introduction to cultural geography*. Rio de Janeiro: Bertrand Brasil, 2003. 224 p.

Geographic itineraries of Rio: free itineraries in defense of the city of Rio de Janeiro. Rio de Janeiro, [20--]. Available at: <http://www.roteirosdorio.com>. Access on: 20 nov. 2015.

RUA, João . Urbanities and New Ruralities in the State of Rio de Janeiro: Some theoretical considerations. In: MARAFON, Gláucio José; RIBEIRO, Marta Foeppel (Org). *Studies of Fluminense Geography*. Rio de Janeiro: Infobook, 2002 A. p. 27-42.

RUA, João. Urbanization in rural areas in the state of Rio de Janeiro In: MARAFON, Gláucio José; RIBEIRO, Marta Foeppel (Org). *Studies of Fluminense Geography*. Rio de Janeiro: Infobook, 2002 B. p. 43-69.

SÁ, Fátima. Burle Marx Not Died. *O Globo Magazine*, Rio de Janeiro, year 5, n. 227, 30 nov. 2008. Irregular pagination.

SACK, Robert David. Magic and space. *Annals ot the association of American geographers*, v. 66, n.2, p. 309-322, 1976.

______. *Human Territoriality*: its theory and history. Cambridge. Cambridge University Press, 1986. 256 p.

SALGUEIRO, Teresa Barata. Spatial and urban temporalities. In: CARLOS, Ana Fani Alessandri; LEMOS, Amália Inês Geraiges (Orgs). *Urban Dilemmas*: new approaches on the city. São Paulo: Context, 2003. 430 p.

SANTOS, Milton. *Space and Method*. São Paulo: Nobel, 1992. 88 p.

______. *Metamorphoses of the Inhabited Space*. 5. ed. São Paulo: Hucitec, 1997. 117 p.

______. *For another globalization*. Rio de Janeiro: Record, 2001.

SANTOS, Milton. *The Nature of Space:* Technique and Time, Reason and Emotion. São Paulo: EDUSP, 2002. 384 p.

SANTOS, Noronha. *The parishes of the old river*. Rio de Janeiro: Editions O Cruzeiro, 1965.

SCHUTZ, Alfred. *Phenomenology and social relations*. Rio de Janeiro: Zahar, 1979. 319 p.

SEAMON, DAVID. Body-subject, time-space routines, and place-ballets. In: BUTTIMER, Anne and SEAMON, David. *The Human Experience of Space and Place*. New York: St. Martin's Press, 1980. p. 148-165.

SEAMON, DAVID. Body-Subject, Space-Time Routines and Dances of Place (translation by Paulo Maurício Rangel). *Geography Magazine*, Niterói, v.3, n.2, p. 4-18, Winter 2013.

SILVA, Armando Corrêa da. *The space out of place*. São Paulo: Hucitec, 1988. 128 p.

SILVA, Kelly Cristina Rodrigues. The memory to think the space: the perspective of the place. *Geography Magazine*, Niteroi, v.5, n.2, p. 26-37, Winter 2015.

SIMMEL, Georg. The Metropolis and Mental Life. In: OLD, Otávio Guilherme. *The Urban Phenomenon*. Rio de Janeiro: Zahar, 1976.

Roberto Burle Max. Rio de Janeiro, [20--]. Available at: <http://sitioburlemarx.blogspot.com.br>. Access on: 20 nov. 2015.

SOARES, Maria Therezinha de Segadas . *Aspects of carioca geography:* Rio de Janeiro: National Council of Geography, 1962.

SOARES, Maria Therezinha de Segadas. The geographical concept of neighborhood and its exemplification in the city of Rio de Janeiro. In: BERNANDES, Lysia; SOARES, Maria Therezinha de Segadas. *Rio de Janeiro - City and Region*. Rio de Janeiro: Biblioteca Carioca, 1990. p. 105-120.

SOUZA, Marcelo Lopes de. The contemporary neighborhood: an essay of political approach. *Revista Brasileira de Geografia*, Rio de Janeiro, p. 139-172, 1989.

______. *Changing the City:* A Critical Introduction to Urban Planning and Management. 3. ed. Rio de Janeiro: Bertrand Brasil, 2004. 556 p.

______. *ABC of Urban Development*. 2.ed. Rio de Janeiro: Bertrand Brasil, 2005. 190 p

TARGINO, Tânia; MONTEIRO, Neide Carvalho. *School Atlas of the City of Rio de Janeiro*. 1. ed. Rio de Janeiro: Municipal Department of Education: Instituto Municipal de Urbanismo Pereira Passos, 2000. 41p.

TUAN, Yu Fu. Topophilia or sudden encounter with landscape. *Landscape*, v.11, n. 1, p. 29-32, 1961.

______. Sacred Space: Exploration of an Idea. In: DIMENSION of human geography. Chicago: University of Chicago, 1978. p. 84-100.

______. *Topophilia*: A Study of Perception, Attitudes and Values of the Environment. São Paulo/Rio de Janeiro: DIFEL, 1980. 288 p.

TUAN, Yu Fu. Humanistic Geography. In: CHRITOFOLETTI, Antônio. *Perspectives of Geography*. São Paulo: DIFEL, 1982. p. 143-164.

______. *Space and Place*: The Perspective of Experience. São Paulo: DIFEL, 1983. 250 p.

______. *The good life*. Madison: The University of Wisconsin Press, 1986. 191 p.

______. A view of geography. *Geographical Review*. New York, v. 81 n. 1: p. 99-106, 1991.

______. *Escapism*. Baltimore: The Johns Hopkins University Press, 1998. 245 p.

______. *Landscapes of Fear*. São Paulo: UNESP, 2005. 373 p.

______. Space, Time and Place: A Humanistic Framework. *Geography Magazine,* Niterói, v.1, n.1, p. 8-19, Winter 2011.

______. *Topophilia:* A Study of Perception, Attitudes and Values of the Environment. Londrina: Eduel, 2012. 344 p.

______. *Space and Place:* The Perspective of Experience: Londrina, PR: EDUEL, 2013. 248 p.

WEID, Elisabeth von der. *The streetcar as an element of urban expansion in Rio de Janeiro.* Rio de Janeiro: Casa de Rui Barbosa Foundation, 1997. 30 p.

WIRTH, Louis. Urbanism as a way of life. In: OLD, Otávio Guilherme. *The Urban Phenomenon.* Rio de Janeiro: Zahar, 1976. p. 90-113.

WRIGHT, John. Terrae incognitae: the place of the imagination in geography. *Annals of the Association of American Geographers,* v. 37. p. 01-15, 1947.

YÁZIGI, Eduardo. Urban Environmental Heritage: remaking a concept for urban planning. In: CARLOS, Ana Fani Alessandri; LEMOS, Amália Inês Geraiges (eds). *Urban Dilemmas*: New Approaches to the City. São Paulo: Context, 2003. p. 253-265.

ABOUT THE AUTHOR

Marcio Luis Fernandes was born on January 25, 1973 in Ilha de Guaratiba, a neighborhood in the city of Rio de Janeiro where he still lives.

Son of housewife Vandi Bastos de Moraes and gravedigger Manoel Fernandes, he attended elementary school at Escola Municipal Narcisa Amália (1980-1989) and high school at Colégio Estadual Freire Alemão (1990-1992).

In 2001, he began his degree in geography at UNIMSB. In 2005, he went to Rio de Janeiro   State University - UERJ - where he specialized in territorial politics in the state of Rio de Janeiro (2005-2006), master's degree (2008-2010) and doctorate (2012-2015) in geography.

Married to teacher Jeane Bazzani and father of Nicole Bazzani Fernandes, he is currently professor of geography in the public networks of the state and city of Rio de Janeiro.

I want morebooks!

Buy your books fast and straightforward online - at one of world's fastest growing online book stores! Environmentally sound due to Print-on-Demand technologies.

Buy your books online at
www.morebooks.shop

Kaufen Sie Ihre Bücher schnell und unkompliziert online – auf einer der am schnellsten wachsenden Buchhandelsplattformen weltweit! Dank Print-On-Demand umwelt- und ressourcenschonend produziert.

Bücher schneller online kaufen
www.morebooks.shop

KS OmniScriptum Publishing
Brivibas gatve 197
LV-1039 Riga, Latvia
Telefax: +371 686 204 55

info@omniscriptum.com
www.omniscriptum.com

Printed by Books on Demand GmbH, Norderstedt / Germany